YOUR KNOWLEDGE HAS VALUE

- We will publish your bachelor's and master's thesis, essays and papers

- Your own eBook and book - sold worldwide in all relevant shops

- Earn money with each sale

Upload your text at www.GRIN.com
and publish for free

Bibliographic information published by the German National Library:

The German National Library lists this publication in the National Bibliography; detailed bibliographic data are available on the Internet at http://dnb.dnb.de .

This book is copyright material and must not be copied, reproduced, transferred, distributed, leased, licensed or publicly performed or used in any way except as specifically permitted in writing by the publishers, as allowed under the terms and conditions under which it was purchased or as strictly permitted by applicable copyright law. Any unauthorized distribution or use of this text may be a direct infringement of the author s and publisher s rights and those responsible may be liable in law accordingly.

Imprint:

Copyright © 2017 GRIN Verlag, Open Publishing GmbH
Print and binding: Books on Demand GmbH, Norderstedt Germany
ISBN: 9783668526051

This book at GRIN:

http://www.grin.com/en/e-book/372819/electron-transmission-through-non-tunneling-regime-of-gaas-algaas-coupled

Sujaul Chowdhury, Mizanur Rahman

Electron transmission through non-tunneling regime of GaAs-AlGaAs Coupled Quantum Wells

GRIN Publishing

GRIN - Your knowledge has value

Since its foundation in 1998, GRIN has specialized in publishing academic texts by students, college teachers and other academics as e-book and printed book. The website www.grin.com is an ideal platform for presenting term papers, final papers, scientific essays, dissertations and specialist books.

Visit us on the internet:

http://www.grin.com/

http://www.facebook.com/grincom

http://www.twitter.com/grin_com

Electron transmission through non-tunneling regime of GaAs-AlGaAs Coupled Quantum Well

Mizanur Rahman, Sujaul Chowdhury*
Nanostructure Physics Computational Lab., Department of Physics,
Shahjalal University of Science and Technology, Sylhet 3114, Bangladesh.

We have investigated electron transmission through non-tunneling regime of a semiconductor nanostructure called Coupled Quantum Well (CQW). Oscillatory transmission coefficient as a function of energy is found to show spectacular waxing and waning in amplitude. Corresponding features are expected to be observed in optical (Physical Review B **54** (1996) 1541) and quantum transport (Physical Review Letters **58** (1987) 816) experiments and hence the results besides verifying the Physics will have impact on future devices based on CQW.

I. Introduction

With development of molecular beam epitaxy, two-dimensional systems similar to Quantum Wells (QWs) have been intensely studied in the past two decades and the results are about confined states in QWs. In recent years, however, people have investigated unbound states[3-23] in a series of semiconductor heterostructures including QWs. G Bastard[3] presented theoretical study of continuum states of QW of "separate confinement heterostructures". G. Bastard et al.[4] studied bound and virtual bound (resonant) levels of GaAs-AlGaAs CQWs and showed parametric variation of energy levels. G. Bastard[5] presented results of calculations of bound and virtual bound states of QWs, "separate confinement heterostructures" and superlattices. M. Heiblum et al.[6] reported theoretical and experimental study of electrons transmission through non-tunneling regime of QW structures and ventured to relate transmission coefficient with I-V characteristics. W. Trzeciakowski et al.[7] presented a simple method of calculating global density of states in resonant tunneling structures. S. Fafard[8] have studied transmission coefficient in non-tunneling regime of QWs. Jian-Ping Peng et al.[9] have studied transmission coefficient in non-tunneling regime of double barrier structures; they presented analytical results of life-time for both quasi-bound and extended resonant states. W. Trzeciakowski et al.[10] have studied change in density of states in above barrier states of symmetric and asymmetric QWs introduced by some localised potential. W. Trzeciakowski et al.[11] have derived dispersion relations connecting transmission amplitude and change of density of states introduced by scattering potential. Marcello Colocci et al.[12] have demonstrated existence of above-barrier states in double barrier QW structures. C. D. Lee et al[13] have reported direct observations of numerous above barrier states in QW structures. W. Lu et al.[14] have demonstrated existence of above-quantum-step quasi-bound states in QW heterostructures. R. Ferreira and G. Bastard[15] have presented an excellent review of unbound states in quantum heterostructures.

Bandgap engineering provides possibility of controlling electronic and optical properties of semiconductor QW heterostructures by varying layer thickness and material compositions. Several non-conventional structures have been proposed and realized for both investigating peculiar aspects of carrier physics and improving performances of QW devices. Two quantities have been frequently used to characterize the properties: transmission coefficient and density of states. Now-a-days people are more interested about QW applications in next generation applications in nanoscale. C. S. Liu[16] investigated exciton dynamics in CQW. Wei Wei et al.[17] conducted a study on TMD QWs and the results indicate that QWs hold promise in wide range of applications. More recent work can be found in reference[18, 19].

S. Chowdhury and C. C. Sarker[20] investigated electron transmission through non-tunneling regime of isolated QW and found oscillations in transmission coefficient versus energy curves monotonic in amplitude. S. Chowdhury and M. Hasan[21] investigated electron transmission through non-tunneling regime of single rectangular tunnel barrier and also found oscillations in transmission coefficient versus energy curves monotonic in amplitude. P. Sutradhar and S. Chowdhury[22] investigated electron transmission through non-tunneling regime of symmetric rectangular double barrier and found spectacular waxing and waning in amplitude of oscillatory transmission coefficient versus energy curves. S. Chowdhury and A. Rahman[23] obtained parametric variations of energy of transmission peaks of symmetric rectangular double barrier in non-tunneling regime and found to their surprise that one of the 3 parametric variations in non-tunneling regime is completely different from that in tunneling regime.

This paper presents a theoretical study of transmission of electron through non-tunneling regime of GaAs-AlGaAs CQW. The paper is organized as follows. In sec. II, we present analytical results on transmission coefficient of CQW. A discussion of the results is reported in sec. III and the concluding remarks are given in Sec. IV.

II. Transmission coefficient of Coupled Quantum Well for non-tunneling regime

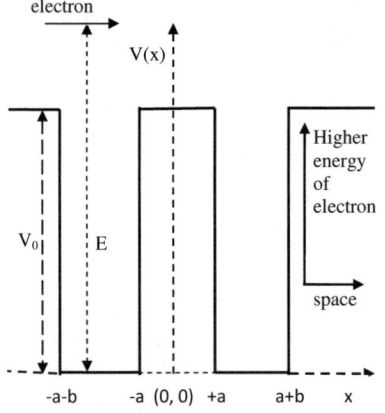

FIG. 1. Band model of GaAs-AlGaAs Coupled Quantum Well in conduction band. V_0 is taken as conduction band offset 773x meV where x is Al content of $Al_xGa_{1-x}As$.

Using the potential energy profile of Fig 1, we have calculated transmission coefficient T(E) of Coupled Quuantum Well for $E > V_0$ i.e. for non-tunneling regime and obtained

$$T = \frac{T_1^2}{T_1^2 + 4(1-T_1)\cos^2[\beta(2a+b)-\theta]} \quad \text{------(1)}$$

where $T_1 = \dfrac{1}{1 + \dfrac{1}{4}\dfrac{V_0^2}{E(E-V_0)}\sin^2\left(b\sqrt{\dfrac{2mE}{\hbar^2}}\right)}$,

$\theta = -\tan^{-1}[\dfrac{1}{2}(\dfrac{\alpha}{\beta}+\dfrac{\beta}{\alpha})\tan\alpha b] + b\beta$,

$\alpha^2 = \dfrac{2mE}{\hbar^2}$,

$\beta^2 = \dfrac{2m(E-V_0)}{\hbar^2}$

Here T_1 is transmission coefficient of single Quantum Well.

III. Results and discussions

We have used the equations of section II in the form of programs written in Mathematica and have plotted T as function of E and investigated characteristic features of the T versus E curves. Typical T versus E curve is shown by undashed curves in Fig. 2(c) and (d); looking at these curves, we find striking amplitude modulated oscillatory behaviour of T versus E curve of Coupled Quantum Well for non-tunneling regime. Amplitude of the oscillations show spectacular waxing and waning as a function of energy E.

Key to understanding the oscillations and the waxing and waning of amplitude of oscillations is equation (1) in which we find that for a given value of T_1, the oscillations result from oscillatory $cos^2[\beta(2a+b)-\theta]$ term which oscillates between 0 and 1. As such T oscillates between 1 and $\dfrac{T_1^2}{T_1^2+4(1-T_1)}$. Maxima of T are always 1 but minima are $T_{min} = \dfrac{T_1^2}{T_1^2+4(1-T_1)}$ which is lower or smaller if T_1 is lower. Thus oscillatory T_1 as function of E causes waxing and waning of amplitude of oscillations of T versus E curve.

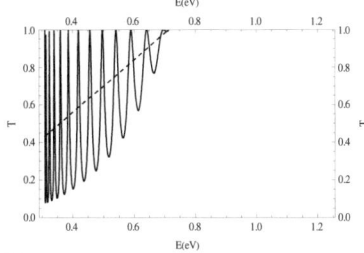

(a)

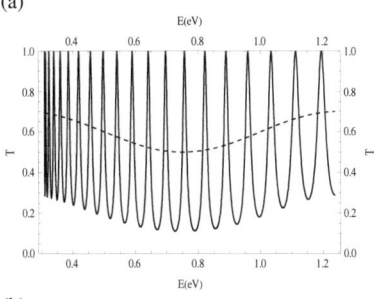

(b)

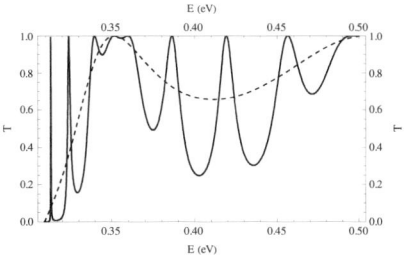

(c)

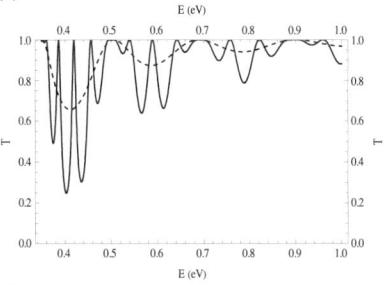

(d)

FIG. 2. Undashed: T versus E curve of Coupled Quantum well for non-tunneling regime. Dashed: T_1 versus E curve of single QW. For QW width b = 20 nm, Tunnel barrier width 2a = 36 nm, Al content of $Al_xGa_{1-x}As$ x = 0.4. Dashed curves in (a) and (b) show T_1 as function of E for model variation of T_1 given by (a) T_1 = 1.4 E, (b) T_1 = 0.6+ 0.1sin($2\pi E$ /1) and dashed curves in (c) and (d) show actual variation of T_1 as a function of E.

We now demonstrate this using model variations of T_1 as function of E: T_1 = 1.4 E for Fig 2(a), and T_1 = 0.6+ 0.1sin($2\pi E$/1) for Fig 2(b). In Figure 2(a), dashed curve shows T_1 rising linearly as a function of E; as such, amplitude of oscillations in undashed T vs E curve wanes rapidly as E rises.

In Figure 2(b), dashed curve shows T_1 reducing from 0.7 to 0.5 before rising again to 0.7 as E rises; as such, amplitude of oscillations in undashed T vs E curve waxes before waning.

We now demonstrate variation of amplitude of oscillations of T versus E curves for actual variations of T_1 as function of E. See Fig 2(c) and (d). In Fig 2(c), as E rises from V_0 to 0.35 eV, dashed curve shows T_1 rising rapidly from 0 to 1; as such undashed curve shows rapid waning in amplitude of T vs E curve in this energy range; compare with Fig 2(a). Between E = 0.35 eV and 0.5 eV in Figure 2(c), T_1 reduces from 1 to 0.65 before rising to 1. As such, as undashed curve in Fig 2(c) shows, oscillations in T vs E curve wax in amplitude before waning; Fig 2(d) shows a series of such waxing and waning in amplitude of T vs E curve.

IV. Conclusions

To conclude, we have found spectacular waxing and waning in transmission coefficient versus energy curves of Coupled Quantum Well for non-tunneling regime. It provides the possibility of controlling the electronic and optical properties of semiconductor Quantum Well heterostructures by varying the layer thickness and material compositions. We believe that the observed waxing and waning in amplitude might be of some relevance towards the realization of novel Quantum Well devices. Corresponding features are expected to be observed in optical[13] and quantum transport[6] experiments and hence the results will have impact on future devices based on Coupled Quantum Well.

Acknowledgements

We are grateful to the Department of Physics, SUST.

References

* s.chowdhury-phy@sust.edu
http://schowdhury-phy.weebly.com
www.sust.edu

[1] Sujaul Chowdhury; "Quantum Mechanics" Narosa / Alpha Science (2014)

[2] Sujaul Chowdhury; "Nanostructure Physics and Microelectronics" Narosa / Alpha Science (2014)

[3] G. Bastard; Physical Review B **30** (1984) 3547.

[4] G. Bastard, U. O. Ziemelis, C. Delalande, M. Voos, A. C. Gossard, W. Wiegmann; Solid State Communications **49** (1984) 671.

[5] G. Bastard; Superlattices and Microstructures, **1** (1985) 265.

[6] M. Heiblum, M. V. Fischetti, W. P. Dumke, D. J. Frank, I. M. Anderson, C. M. Knoedler, L. Osterling; Physical Review Letters **58** (1987) 816

[7] W. Trzeciakowski, D. Sahu, Thomas F. George; Physical Review B **40** (1989) 6058.

[8] S. Fafard; Physical Review B **46** (1992) 4659

[9] Jian-Ping Peng, Yao-Ming Mu, and Xue-Chu Shen; Journal of Applied Physics **73** (1993) 989.

[10] W. Trzeciakowski and M. Gurioli; Journal of Physics: Condensed Matter **5** (1993) 105.

[11] W. Trzeciakowski, M. Gurioli; Journal of Physics: Condensed Matter **5** (1993) 1701.

[12] Marcello Colocci, Juan Martinez-Pastor, Massimo Gurioli; Physical Review B **48** (1993) 8089.

[13] C. D. Lee, J. S. Son, J. Y. Leem, S. K. Noh, Kyu-Seok Lee, C. Lee, I. S. Hwang, H. Y. Park; Physical Review B **54** (1996) 1541.

[14] W. Lu, Y. M. Mu, X. Q. Liu, X. S. Chen, M. F. Wan, G. L. Shi, Y. M. Qiao, S. C. Shen, Y. Fu and M. Willander; Physical Review B **57** (1998) 9787.

[15] R. Ferreira and G. Bastard; Nanoscale Research Letters **1** (2006) 1:120–136.

[16] C. S. Liu, H. G. Luo and W. C. Wu, Physical Review B **80** (2009) 125317

[17] W. Wei, Y. Dai, C. Niu, B. Huang, Scientific reports **5**, Nature 17578 (2015).

[18] J. Frigerio, V. Vakarin, P. Chaisakul, M. Ferretto, D. Chrastina, X. Le Roux, L. Vivien, G. Isella, D. Marris-Morini, Scientific Reports **5**, Nature 17398 (2015).

[19] Ahmed Fadil, Daisuke Iida, Yuntian Chen, Jun Ma, Yiyu Ou, Haiyan Ou, Scientific Reports **4**, Nature 6392 (2014).

[20] S. Chowdhury and C. C. Sarker; Oscillatory transmission through non-tunneling regime of isolated Quantum Well, Grin Verlag (2013)

[21] S. Chowdhury and M. Hasan; Oscillatory transmission through non-tunneling regime of single rectangular tunnel barrier, Grin Verlag (2013)

[22] P. Sutradhar and S. Chowdhury; Oscillatory resonant transmission waxing and waning in amplitude, Lambert Academic Publishing (2012)

[23] S. Chowdhury and A. Rahman; Oscillatory Transmission Through Non-Tunneling Regime of Symmetric Rectangular Double Barrier, GRIN Verlag (2013)

YOUR KNOWLEDGE HAS VALUE

- We will publish your bachelor's and master's thesis, essays and papers

- Your own eBook and book - sold worldwide in all relevant shops

- Earn money with each sale

Upload your text at www.GRIN.com
and publish for free